FOUNDATIONS OF PHYSICS

SECOND EDITION

MASSIMO F. BERTINO
ROBERT H. GOWDY
Virginia Commonwealth University

Kendall Hunt
publishing company

Kendall Hunt
publishing company

www.kendallhunt.com
Send all inquiries to:
4050 Westmark Drive
Dubuque, IA 52004-1840

Printed in the United States of America
10   9   8   7   6   5   4   3   2

Contents

CHAPTER 1

What Is Science?

Science is closely related to human nature. It arises from the spirit of inquiry of human beings. You see something new, be it a planet, a rainbow or an IPod, and you start wondering how it works, where it comes from and what it could be good for.

In science, we observe, make assumptions, and then go back to observation to verify assumptions.

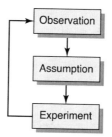

The scientific activity differs from other human endeavors mostly because of the circular step in the figure above. That is, assumptions made to understand a certain process must be validated by additional observations. If the new observations fall in line with our expectations, we move on. Otherwise we ditch our assumptions and look for another explanation.

Example

I want to understand how things fall. I start dropping objects. After a few trials I notice that a plastic sphere and a metal sphere of the same radius fall in the same way. Then I make an assumption: "Maybe the rate of fall does not depend on the material making up the sphere." To test the assumption, I prepare spheres of the same radius made of different materials. Then I measure how they fall. If these spheres fall in the same way I can move on. If some materials fall differently from others, I have to come up with some other explanation.

This same example shows another important characteristic of science. Science can only be proven wrong, not right. What does this mean? Go back to the example. I prepare spheres out of, say, 10 materials. If even one material behaves differently (say, it starts flying instead of falling), then my assumptions are wrong. However, even if all spheres fall in the same way, a naysayer could argue that I have not proven anything. For example, one could say that some materials available only on Pluto start flying once dropped. Therefore, scientific conclusions are never 100% sure. Recognizing this issue, Popper came up with a definition of scientific statements that is perhaps counterintuitive but sound. "A scientific statement is one that can be proven wrong."

Let's look back at the example of the falling objects. I make a statement, "Spheres of the same radius fall at the same rate, no matter what they are made of." Is this a scientific statement (in Popper's sense)? Yes. It can be proven wrong. You might find a material that behaves differently (say, it starts flying, or it falls at a slower rate) and prove the statement to be wrong. However, does my statement guarantee 100% that all materials will behave the same way? No. Scientific statements (again, in Popper's sense) are required to be tentative and disprovable. At best, they tell us what has not been shown to be wrong.

Popper's definition may sound a bit negative, since we often perceive scientific statements to have an absolute character, and we often call them laws. Where is the discrepancy between Popper's definition and our perception? The difference is in the number of times that a statement has been scrutinized. How many objects made of different materials are being dropped every day? How many of these objects start flying, or stop in mid-air? Based on our experience what are the chances that on Pluto there are objects that will fall differently?

So, scientific evidence is <u>always</u> circumstantial. This does not mean that science is useless or that it is a dream. Without science, we would not have antibiotics, computers or airplanes. Also, some scientific statements have withstood rigorous and often ferocious analysis for more than 300 years. Are we 100% sure that these statements are correct? No. However, betting against these statements is generally not a good idea.

Sometimes, a well-tested scientific theory is proven "wrong". How? By experiment. Does it mean the people who developed and used the theory were idiots? No. These people made a number of experiments to test the theory, and the tests agreed with the theory. Usually, if a theory has been verified and accepted for a long time, new experiments show that the theory is a special case of a more comprehensive phenomenon.

Example

The theory of gravitation. Developed in the 1600s, it remained THE theory until the early 1900s, when Einstein showed that it was a special case of a more general theory. Newton's theory works well when the masses of the objects are small. Einstein showed that for large masses (e.g., a star), other rules apply. Einstein's theory was confirmed by experiment (i.e., not proven wrong) by Eddington.

So, science is nice and fine. We should not forget, however, that while scientific statements are powerful, there are many things in life that are as powerful, yet they cannot be tentative.

Examples

Politics, law Matters of life and liberty cannot be based on tentative decisions. The constitution cannot be proven wrong. It can be amended, though.

In **mathematics and logic**, definitions, axioms, and proven theorems are *not falsifiable*. The content of established mathematics and logic is *not science* (as defined by Popper).

Religion Both testaments make it very clear that the Bible is not about tentative, testable statements. The same is true for the other religions. They are not tentative. Does it mean that religion is wrong, or bad? No, it just means that the criteria that we use in science do not apply to religion.

Art The meaning which you find in a work of art can never be wrong. One may or may not like it, but nobody who is sane of mind can deny the value of art.

We will spend now a little time discussing pseudo-science, which is being often propagated on internet sites.

Pseudo-science

... is something that presents itself like science, looks like science, sounds like science, but it is not science...

Examples

- **Homeopathy.** Look up discussion on quackwatch.com. Homeopathy has been tested several times, and proven not to work each time. Yet, it has been around for more than 100 years.

- Claims that the earth has been visited by **aliens.**

How Do I Discriminate between Science and Pseudo Science?

Pseudo science is often characterized by exaggerated claims, (this herb can cure any cancer...), denial of established procedures (they do not want you to know...the tests disproving our claims were rigged...), lack of equations and clearly verifiable claims, and it often contains a wild mixture of statements.

CHAPTER 2

Describing Motion

In physics, we want to learn why and how things work. Alas, "things" are often very complicated. Therefore, we start often from the study of simple systems, such as objects moving on a straight line. Before starting, however, we must agree on how to measure things. Even a simple statement like "The distance covered by running for 10 minutes", for example, is vague. An average person and an athlete run at different paces. Similarly, "A distance of ten paces" is ambiguous. Ten paces of a tall person are not the same as ten paces of a child. To eliminate these ambiguities, over time most countries established units of length, time, mass, etc., and precise ways of measuring these quantities. Groups of countries then agreed to use the same units to facilitate commerce and information exchange. Today, most countries (regrettably, not the US) agree to measure

Lengths in meters (symbol: m)

Masses in kilograms (symbol: kg)

Time in seconds (symbol: s)

The definitions of most of these units are arbitrary; essentially, people agreed that a chunk of matter would be called a kilogram. We also observe that we can measure masses, length, and time, but we cannot define them. Try to define time, for example. . .

To describe motion, another preliminary thing that we must do is to define a starting point and a positive direction of motion. On highways, the government has done this for us. For example, drive on I-64 at the border between Virginia and West Virginia, and move east. Read the mile numbers:

Border	Lexington	Richmond	Norfolk
Mile 0	mile 55	mile 190	mile 284

Origin   positive x-direction

The government has decided that the origin (i.e. mile 0) is at the border between states and that the positive direction is eastward. This decision is arbitrary (it could have been the other way around), but it is necessary, since it removes any ambiguities. In laboratory experiments we do the same, objects moving along rails with reference marks.

To describe motion, we also use relative numbers, i.e., positive and negative.

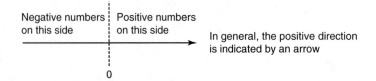

Negative numbers on this side | Positive numbers on this side

In general, the positive direction is indicated by an arrow

0

Important: the positive direction is only a convention
I can also use a positive axis that goes the other way...

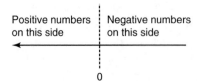

Positive numbers on this side | Negative numbers on this side

0

Example

Define Richmond as the starting point on I-64.
Where is Lexington?
Where is Norfolk?

WVA ┊ VA

LEX	RIC	NOR
−135	0	94

−190

What has changed? The numbers.
Is Lexington still there? Yes, it did not move.
It is just described by a negative number.

Are the distances still the same? Yes.
Before: distance = RIC − LEX = 190 − 55 = 135 miles
Now: distance = RIC − LEX = 0 − (−135) = 135 miles

The same ideas apply to objects moving in the vertical direction:

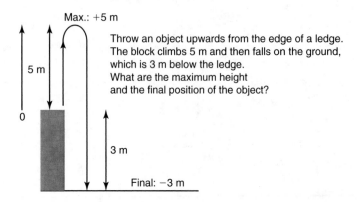

Max.: +5 m

Throw an object upwards from the edge of a ledge.
The block climbs 5 m and then falls on the ground,
which is 3 m below the ledge.
What are the maximum height
and the final position of the object?

5 m

0

3 m

Final: −3 m

Times can also be negative.

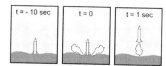

Negative times are those before a certain event; positive times are after the event.

The number of distances needed to fix the position of an object is called the *dimension* of the motion.

In one dimension (e.g., cart moving on a straight road)
I need only ne number to describe the motion.
That number is the distance covered by the object while moving along the line.
The number will be positive if the object moves along the positive axis, negative if it moves opposite to the positive axis.

In two dimensions, the object moves along two directions.

Example

A marble moving horizontally and falling from a desk. The object changes height, but it also moves away from the edge of the desk. Two numbers are needed (horizontal and vertical position) to describe this motion.

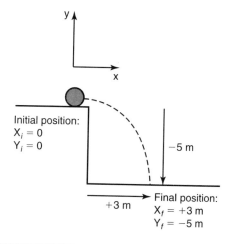

Change and Rate of Change

When things move, they change their position. In a process (not only motion) we have an initial quantity, Q_i, which evolves to a final value, Q_f. The change is defined as

$$\Delta Q = Q_f - Q_i$$

Example

A car is 10 miles from Richmond at 1.00 P.M.

and 100 miles from Richmond at 3.00 P.M.

The *change in distance* is (final value − initial vlaue) or (100 miles − 10 miles) or 90 miles.

We conclude that if the final value is *larger* than the initial value, then the *change* is a positive number.

If the final value is smaller than the initial value, then the change is a negative number.

Example

I pay $100 for a new ipod.

Three months later, I resell it for $50.

What is the change?

$\Delta Q = Q_f - Q_i = 50 - 100 = -50$\$

Does it mean the change is not for real?

No. It means that the change is negative.

The *average rate of change* of a quantity is the change in the quantity divided by the corresponding change in the time.

Initial time = t_i	Initial value = Q_i
Final time = t_f	Final value = Q_f
$\Delta t = t_f - t_i$	$\Delta Q = Q_f - Q_i$

$$\text{Rate of change} = \frac{\Delta Q}{\Delta t}$$

Velocity

Velocity is the rate of change of position.
That is: How long it takes to get that far.

Example

Last night I drove to Norfolk. It took me 2 hours to get there.

What was my average velocity?

First, let's make a sketch . . .

WVA ⋮ VA

LEX   RIC   NOR
55   190   284
             t = 0   t = 2 h

0

- Distance traveled = final − initial = 284 − 190 = 94 miles
- Time traveled = final − initial = 2 h − 0 h = 2 h
- Velocity = distance/time = 94 miles/2 h = 47 miles/h

Do not forget: velocity = space/time

$$V = \Delta s / \Delta t$$

Can velocity be negative?
Yes.

Assume that an object is released from rest from a ledge, and drops on a rock 5 m below the ledge. The object takes 1s to fall. What is its average velocity?

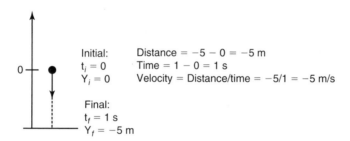

Initial: Distance = −5 − 0 = −5 m
$t_i = 0$ Time = 1 − 0 = 1 s
$Y_i = 0$ Velocity = Distance/time = −5/1 = −5 m/s

Final:
$t_f = 1$ s
$Y_f = −5$ m

Note: this value is negative because the positive vertical axis points upwards. Had the axis pointed downwards, the value would have been positive.

Example

A runner moves along the positive x-direction and covers 100 m in 10 s.
What is its average velocity?

$X_i = 0$ $X_f = 100$
$t_i = 0$ $t_f = 10$

Distance = 100 − 0 = 100 m
Time = 10 − 0 = 10 s
Velocity = 100/10 = 10 m/s

So, man's highest velocity is around 10 m/s. How does that compare to a car?
Since the velocity of cars is usually measured in miles/hour, we have to convert units.
How do we do this?
Let's go first from m/s to km/h

How many m in one km? 1000. That is, I have to divide by 1000.
How many seconds in one hour? 3600. Divide by 3600.

$$10\frac{m}{s} = \frac{10\frac{1}{1000}\,km}{\frac{hr}{3600}} = \frac{10 \times 3600}{1000}\frac{km}{h} = 36\frac{km}{h}$$

So, a person can run at about 36 km/h. Since a mile is about 1.6 km, we obtain that a person can run a tad faster than 20 miles/hr.

Compare with other velocities:

Cheetah ~ 130 km/h

Elephant ~ 40 km/h

Soccer (free kick) ~ 90−150 km/h

Revolution of earth around Sun ~ 107,000 km/h

Light ~ 300,000 km/s (second!)

Before we close with velocity, we introduce the definition of speed. **Speed is the absolute value of velocity.**

Acceleration

Velocity can also change with time. Cars are at rest at a red traffic light, and start moving when the light turns green. Can we quantify the rate of change of velocity? Is it useful?

We define acceleration as the rate of change of velocity.

That is:

a = $\Delta v / \Delta t$

Not an intuitive quantity.

Where do I use the concept of acceleration?

When I compare a pick-up truck with a sports car.

e.g.:

F150	0–60 mi/h in 10 sec
Dodge Viper	0–60 mi/h in 4 sec

Examples

A car is traveling away from Richmond at 10 meters per second.
Five seconds later, it is traveling away from Richmond at 30 meters per second.

What is the component of acceleration away from Richmond? Divide the change in velocity by the time taken:

$$a = (V_f - V_i)/\Delta t$$

$$\frac{30\ m/s - 10\ m/s}{5\ s} = \frac{20\ m/s}{5\ s}$$

$$= \frac{20}{5} \times \frac{m/s}{s} = 4\ m/s^2$$

Notice: m/s^2

Why? $a = v/t = (m/s)/s = m/s^2$

A car is traveling away from Richmond at 30 meters per second.
Five seconds later, it is travling away from Richmond at 10 meters per second.
What is the component of acceleration away from Richmond?
Divide change in velocity by time taken:

$$\frac{10 \text{ m/s} - 30 \text{ m/s}}{5 \text{ s}} = \frac{-20 \text{ m/s}}{5 \text{ s}}$$

$$= \frac{20}{5} \times \frac{\text{m/s}}{\text{s}} = -4 \text{ m/s}^2$$

Acceleration can be negative!

For each distance measurement there is a velocity component and an acceleration component.

So long as distances are always measured in the same directions, the *acceleration component* is just the rate of change of the velocity component.

Note: change in velocity, not speed

An airplane's rate of climb is the rate of increase in its distance above sea-level. Suppose that an airplane is pulling out of a dive.

Its rate of climb changes from -10 m/s to $+10$ m/s in one second.

What is its vertical component of acceleration?

(A) -10 m/s^2

(B) $+10$ m/s^2

(C) $+20$ m/s^2

(D) -20 m/s^2

(E) 0

CHAPTER 3

Falling Objects

Take an object in your hand. Let it go. It falls (duh!). Now think at what happens. In the beginning, $V_i = 0$. After a time t, $V_f \neq 0$. So, there is a change in velocity: $\Delta V = V_f - V_i = V_f$.

This change takes place in a certain time t. The rate of change of the velocity (that is, the acceleration) is

$$a = \frac{\Delta V}{t} = \frac{V_f}{t}$$

By observing a bit more carefully, you notice that the velocity of your object keeps increasing with time. That is, falling bodies are subject to an acceleration throughout their fall, not only at the beginning. These observations were made very early in the history of mankind. Another observation was made, however, which prevented a thorough understanding of falling bodies for a long time. Namely, people dropped a heavy object, like a hammer, and saw that it did fall faster than a light object such as a feather:

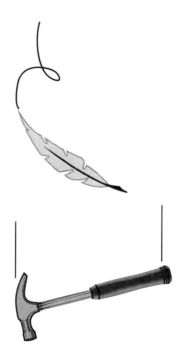

Aristotle reasoned that the hammer falls faster because it is attracted more strongly by the Earth. The explanation made sense (sort of, at least) and it was accepted. Galileo (and others pretty much at the same time) questioned this interpretation. They devised a series of experiments which showed that the reasoning was not correct.

Example

Drop two objects at the same time, both fairly heavy so that air resistance can be neglected.

Aristotle would say that the heavier object should reach the ground **long** before the lighter one.

Actually, they hit at about the same time.

Prepare two identical metal spheres. Cut one in half.

Drop the two halves and the whole sphere at the same time.

They will reach the ground at the same time.

These experiments showed that Aristotle's reasoning was not correct. We also notice a fundamental difference between Galileo and Aristotle, which is one of the characteristics of science. Galileo, like Aristotle, observes reality and tries to understand how things work. However, Galileo makes experiments to prove his conjectures. Aristotle jumps from observations to conclusions, but he does not perform actual tests to prove or disprove his ideas.

Galileo proposed that an explanation (e.g., rate of fall proportional to weight) should be used to make a definite prediction, which should then be *tested*.

If the prediction fails, then the explanation is *wrong*.

So, why were Aristotle's ideas so accepted?
Because they made sense.

Why are Galileo's ideas superior?
Because they also made sense, but, most importantly,
because **experiments** proved Aristotle's ideas to be wrong,
while not disproving Galileo's ideas.

After many experiments, Galileo came to the conclusion that

**To the extent that friction can be ignored,
all objects fall alike.**

Today, we know that near the surface of the Earth,
objects fall with a constant acceleration vector, which we call g
The vector g points downward and
has magnitude <u>close</u> to 10 m/s².

What does this mean?

It means that for every second an object is in mid-air,
the speed of an object

increases by 10 m/s when the object is moving **down**

decreases by 10 m/s when the object is moving **up.**

Example

Suppose that a brick falls from the top of a building and that it takes
4 seconds to reach the ground. How fast will it be going when it hits?

It gains 10 m/s for every second that it falls, so it will be going
4 × 10 m/s or 40 m/s when it hits.

Free−fall in one dimension

Let's first revisit the definition of velocity and acceleration:

$$v = \Delta x / \Delta t$$

$$a = \Delta v / \Delta t$$

What does this really mean?
First off, let us rewrite:

$$\Delta s = v\Delta t, \text{ that is: } X_f - X_i = v\Delta t$$

$$\Delta x = a\Delta t, \text{ that is: } V_f - V_i = a\Delta t$$

Examples

1. A car travels at 10 m/s.
   After 1s, it has covered $x = v\Delta t = 10$ m.
   After 2 s, it covers 20 m. And so on . . .

2. A car starts from rest with a positive acceleration of 10 m/s².
   In the beginning, $v = 0$.
   After 1 s, $v = 0 + a\Delta t = 10 \times 1 = 10$ m/s.
   After 2 s, $v = 20$ m/s and so on . . .

3. A car has a positive velocity $v = 30$ m/s, and a negative acceleration
   $a = -10$ m/s².
   After 1 s, $v = 30 + a \Delta t = 30 + (-10) \times 1 = 20$ m/s (it slows down!)
   After 2 s, $v = 10$ m/s.
   After 3 s, $v = 0$.
   After 4 s, $v = 30 - 10 \times 4 = -10$ m/s. (i.e., now it goes in the opposite direction)

The last example tells us something important and can be revisited with a cartoon.
The car in the picture below starts to spin its tires in reverse while moving forward.
It comes to a stop and then moves backwards.
The velocity vector is red. The acceleration vector is yellow.

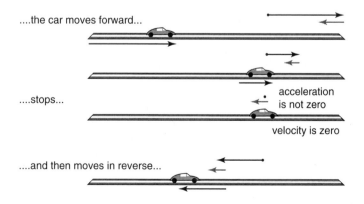

....the car moves forward...

....stops...

acceleration is not zero

velocity is zero

....and then moves in reverse...

Now, let's turn these examples by 90°, i.e., from the horizontal to the vertical. An object moving along the vertical direction will have an acceleration of magnitude 10 m/s² pointing downwards. This means that the speed of an object will increase each second by 10 m/s if the object is moving downward. The velocity will <u>decrease</u> by 10 m/s if the object is moving upwards. The more interesting case is when things are thrown upwards. The slowing down imposed by g means that the object will stop at some point in mid-air.

Fact: An object thrown upward reaches the highest point of its trajectory when its velocity becomes zero (for just an instant).

Why?
Its initial upward velocity keeps decreasing because of g until it stops

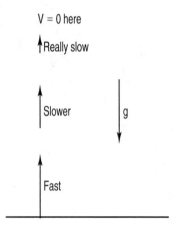

Example

Suppose that an object is thrown upwards at a speed of 30 m/s.
How long will it take to reach the highest point of its trajectory?

It loses 10 m/s every second.
In 3 seconds it will lose 30 m/s and stop.
So, the time to reach the highest point is 3 seconds.

Fact: Many of these problems are symmetric!

Imagine watching a movie of an object thrown upward, losing *10* m/s of speed every second.

Now imagine that the movie is running backwards. Now you see an object falling downward, gaining *10* m/s every second.

i.e., the upward trajectory is just the same as the downward one, but running backwards.

Fact: If an object is thrown upward, the time for it to come back down to the same level is twice the time it takes to reach its highest point.

Example

An object is thrown upwards at a speed of 20 m/s.
How many seconds will it take to come back down?
The time to reach its highest point is 2 seconds—the time needed to lose 20 m/s of speed.

The downward trajectory is exactly the upward one played backwards and will take the same amount of time—2 seconds.
Total time = 2 s + 2 s = 4 s.

Guess what happens to the velocity. . . .

An object is thrown upwards with a speed of 30 m/s.

How fast will it be going when it comes back down?

Since the downward trip is just the upward trip played backwards, it will be going 30 m/s downward.

Alternatively: it takes 3 seconds to get to the top of the trajectory.

When on top, the object has zero velocity and starts falling.

The problem is symmetric, so it takes 3 seconds to get down.

In these 3 seconds, its velocity will increase and reach the final value of $3 \times 10 = 30$ m/s.

Careful: the final velocity will point down. . . .

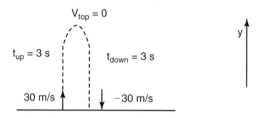

Motion in 2- and 3-D

Motion in a gravitational field in 2- or 3-dimensions is more complicated. It is important to understand, however, because objects <u>do</u> move in more than one dimension in real life. In the 1600s, the theory was developed to understand the motion of cannonballs, a cutting-edge issue at the time.

Since this is complicated, let's simplify . . .

Consider the object that moves horizontally on a table and then falls on the ground.

Experimentally, we find that
- the horizontal component remains the same.
- the initial vertical velocity is zero.
- the final vertical velocity is non -zero and points down.

Why?
Because the horizontal acceleration is zero.
The vertical acceleration points down.

In the more general case (e.g. gun shooting at an angle ϑ from the horizontal) we find that
  1. The trajectory is parabolic.
  2. The motion along the horizontal axis is decoupled from the motion along the vertical axis.

3.   There is no horizontal acceleration, i.e., the horizontal velocity is constant.
4.   The vertical motion follows the laws of one-dimensional vertical motion that we saw earlier.
5.   The maximum range is attained when $\vartheta = 45°$

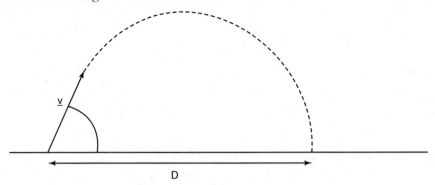

Let us now analyze the case of an airplane dropping a bomb.

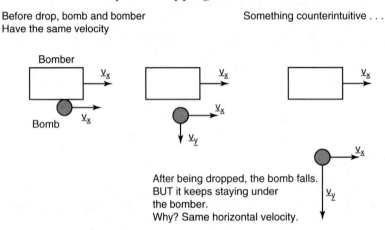

This is why nuclear bomber pilots reverse course as soon as they release their bombs. . .

CHAPTER 4

Why Do Things Move?

First of all, we have to revisit one of the fundamental findings of Galileo's theory of 2-D motion in a gravitational field: the horizontal velocity remains constant.

WHY?

The question had been (once again) asked by Aristotle, who concluded that things, when left alone, stop moving. For example, if you stop pushing on a cart, it stops (sooner or later). However, there were (and are) exceptions to this conclusion. Most notably (in ancient times), arrows. Arrows do not stop moving for a long time after being released from a bow. Aristotle (and others after him) did not have a convincing explanation for this. Galileo thought that things, once they start moving, want to keep moving on a straight line and at a constant speed. Different from Aristotle, Galileo designed experiments to verify his ideas. For example, he made objects moving on inclined planes.

Proof

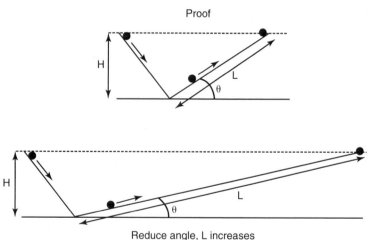

Reduce angle, L increases

Galileo noticed that the distance L covered on the second inclined plane increased when the angle decreased. When the angle was zero, the object kept moving forever. He concluded that bodies, when left alone, remain at rest OR they keep moving at a constant velocity. We call this property of bodies inertia. Was this ever contemplated by Aristotle? No. Are we 100% sure that the law of inertia really applies to all bodies? No, but experiments have been carried out for more than 300 years, and all agree with the law of inertia. The odds of finding a body not obeying the law of inertia are not very large.

Now, Newton started from Galileo's law of inertia and made additional considerations.

Observation: to move things, I need a force. (Duh!)

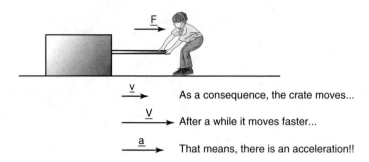

So, a force induces an acceleration.
The crate changes velocity (i.e., moves)
because of the acceleration.

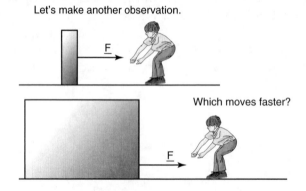

CONCLUSION

Things move because forces are applied.
The acceleration is bigger when the mass is smaller . . .

a = F/m

or, better:

$$F = ma$$

Mass (again!)

The quantity that measures how difficult it is to change the state of motion of an object is called the *mass* of the object.

> When interacting with other objects,
> an object with a *large mass*
> will accelerate *less*
> than an object with a *smaller mass*.

Important: the mass of an object is the same everywhere.
That is, an object with mass m on Earth will have the same m on the Moon.

So, things move because

1. A force is applied.
2. The force causes an acceleration.
3. Because of the acceleration, the velocity changes.

There is a relationship between force, mass and acceleration

$$F = ma$$

Meaning: if I apply the same force to two different objects, the one with a smaller mass will accelerate more (i.e., move faster)

Question: Does this description always hold?

Say I sit on a bench holding a glass of water.
A friend of mine is sitting on a bus and holds a glass of water.
For both of us, the water is at rest inside the glass.
That is, the system that we are observing behaves in the same way.

Now, assume that the bus stops very rapidly.
What happens?
Nothing to the water in my glass.
The water in my friend's glass, instead, starts moving; waves appear inside the glass.
It is as if someone were shaking the glass.
That is, the description of a physical system changes when the observer is not moving at constant velocity.

Let's see what is happening, and why.

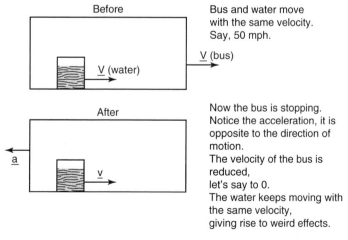

Before

Bus and water move with the same velocity. Say, 50 mph.

$\underline{V}$ (bus)

$\underline{V}$ (water)

After

Now the bus is stopping. Notice the acceleration, it is opposite to the direction of motion.
The velocity of the bus is reduced,
let's say to 0.
The water keeps moving with the same velocity,
giving rise to weird effects.

$\underline{a}$

v

Let's put things together.
Newton's laws (for now)

1. **Law of inertia: an object not subject to external forces will remain at rest or keep moving at constant velocity.**

2. **F = ma**

Important: systems can be described by these laws only in inertial reference frames.

An inertial reference frame is any observer (and laboratory) that is at rest, or moving at constant velocity.
Once again, velocity, not speed.

The laws of physics are the same for all observers in inertial frames. In other words, observers in inertial frames will describe the SAME process in the SAME way.

Why velocity and not speed?

Assume a car is taking a sharp right-angle turn.
Assume it keeps moving at, say, 50 mph through the whole process.
The magnitude of the velocity, i.e., its speed, remains constant (50 mph).
However, the direction changes. That is, the velocity changes.
(reminder: vectors have a length, but also a direction).

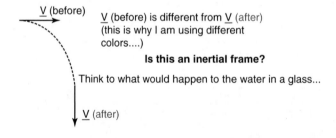

$\underline{V}$ (before) is different from $\underline{V}$ (after)
(this is why I am using different colors....)

Is this an inertial frame?

Think to what would happen to the water in a glass...

Concluding remark: Units of force. They are weird . . .
m = kg
a = m/s²
F = ma = kg m/s²
We call this thing a Newton (N)

Some examples of forces

Weight. Is it a force? Put an anvil on your chest and you will know.
Where does it come from?
I know that all bodies fall with the same acceleration, **g.**

Since there is an acceleration, there must be a force . . . .
I call this force weight and indicate it with **W.**

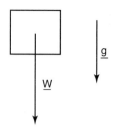

Examples of weight

Consider a mass m =1 kg. What is its weight?
I know that the mass, when left alone, would fall with an acceleration
which has a magnitude close to 10 m/s². The weight is

W =1 × 10 =10 N.

So, 10 N is the force with which Earth's gravitation attracts a mass of 1 kg.

Let's work the other way around . . .

A mass has a weight of 1 N. That is, Earth exerts a force of 1 N on this mass. How big is the
mass?

1. 1 kg.
2. 10 kg.
3. 0.1 kg.

Let's compare with known things

Typical mass of a person: 150 lb.
That makes 70 kg (give or take as usual).
What is the weight?
Misconception: weight is not mass!
W = mg = 70 × 10 = 700 N.

That is, the weight of a person with a mass of 70 kg is 700 N.

How big is a N?

Assume an object has a weight of 1 N.
What is the mass?
Since W = mg,   m = W/g = 1/10 = 0.1 kg ~ 0.2 lb

So, 1 N is not a particularly big force.

Also, we notice that weight depends on g.

> How can I change g?
> Not really easy, I have to change planet

Find the weight in Newtons of a 100 kilogram football player on the Moon (where the acceleration due to gravity on the surfac e is about 1.5 m/s).

$W = mg = (100 \text{ kg})(1.5 \text{m/s}) = 150\text{N}$

Notice that this is a lot less than his weight on Earth:

$W = mg = (100 \text{ kg})(10 \text{m/s}) = 1000\text{N}$.

Your *weight* depends on where you are. Your mass does not.

Normal force

Consider the system in the figure:

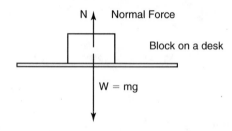

The block does not fall, therefore there must be another force supporting the block. We call this force the normal force and indicate it by N.

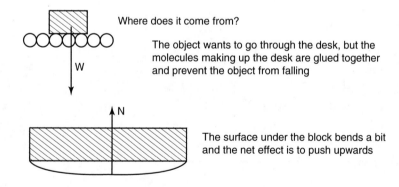

The normal force is always perpendicular to the surface.
That is, on a horizontal surface the normal force is vertical and points up

Where does it come from?

The object wants to go through the desk, but the molecules making up the desk are glued together and prevent the object from falling

The surface under the block bends a bit and the net effect is to push upwards

---

Example

Place a block with m = 10 kg on a desk. The block remains at rest on the desk.
What is the magnitude of the normal force?
The normal force is equal to the weight, only points the other way

N ⬆   So, N = mg = 10 × 10 = 100 N

W = mg

Wait a moment . . . Does this mean the normal depends on the block?

Yepp. . . .

When the mass is bigger the desk bends a bit more, and the force is bigger.

If you do not believe it, start putting weights on a thin wooden rod.

The rod bends more when the load is increased (and eventually breaks).

Force of friction

Friction, at least at an intuitive level, is related to the stopping of objects. To understand this, let's consider a block sliding on a horizontal surface. It starts with a velocity v, say, pointing along the positive x axis, and it stops after a while.

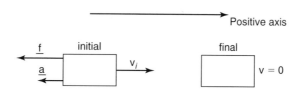

Positive axis

Velocity change is positive or negative?

Where is the force pointing?

This is the force of friction

The velocity changes, therefore there is an acceleration. The acceleration points in the negative direction:

$$a = \frac{0 - v_i}{\Delta t} = \frac{-v_t}{\Delta t}$$

Since there is an acceleration, there must be a force, and the force will be parallel to the acceleration. This force is the force of friction.

Where does friction come from?

Surfaces are rough, even flat ones

Remember, friction is a force that always opposes motion

It turns out that the magnitude of friction is proportional to the weight.
Better said, friction is proportional to the normal force.

$f = \mu N$

Why? More weight will press those rough spots on the surfaces closer
and stronger together.

μ is called coefficient of friction, and it depends on the material.

Example

A block is sliding on a rough surface with coefficient of friction $\mu = 0.1$
The mass of the block is m = 10 kg, calculate the magnitude of the force of friction.

I know that $f = \mu N$. I need N.
I know that N = weight = mg.
Therefore: N = mg = 10 x 10 = 100 N.
So, $f = 0.1 \times 100 = 10$ N.

Tension

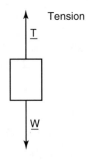

Tension

Tie a rope to the ceiling, hang a block to it.
The block has a mass, gravitation will attract it.
i.e., there is a weight pulling down . . .
The block does not fall, therefore there must be an upward force equal and opposite to the
weight.
The force is exerted by the rope, and we call this force tension, T.

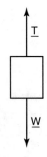

For a block hanging from a rope in an inertial frame of reference,
T = W = mg. Once again, magnitude.

This is how some scales work.
They measure the tension to measure the weight.

Now, assume that a rope is holding a block.
If the weight does not move,
T = W = mg, and a = 0.

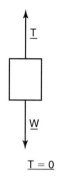

Now cut the rope. T = 0, and
the block will fall with acceleration
g = 10 m/s²

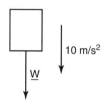

Now, let's go to an intermediate case.
We lower down the block with a constant acceleration.
The magnitude of this acceleration is less than 10 m/s².
Call the magnitude of this acceleration a.
What happens to T?
When a = 0, T = mg.
When a = g, T = 0.

So, T will be smaller than mg, but larger than zero.

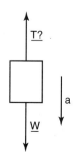

Now, let's make the block go up with constant
acceleration a.

What happens to T?

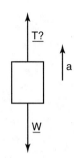

       a. T is zero.
       b. $T = mg$.
       c. T is larger than mg.
       d. The problem cannot be solved.

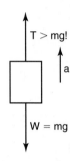

Explanation: put a short-sighted bug sitting on the block.

The bug sees only the rope from which the block hangs.

Since the block is moving upwards, this means that the rope is pulling up with a force larger than the
weight. (if the force were equal to the weight the block would remain at rest).

Now let's put the block on the floor of an elevator.

When the elevator is at rest, the floor exerts an upward norm force equal to the weight.

When the elevator accelerates upwards, the normal force is larger than the weight.

When it accelerates downwards, the force is smaller.

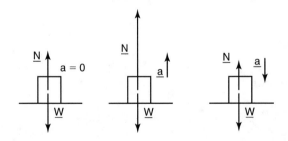

A final remark:

throughout the discussion, we assumed that applying two or more forces
to an object was legitimate.

Actually, it is legitimate.

That is, if I have more forces:

$\underline{F}_1 + \underline{F}_2 + \underline{F}_3 + \ldots = m\underline{a}$ (note: all vectors here, not magnitudes).

We call this thing principle of **superposition of forces.**

Why is superposition so obvious?

Well, if two people push on the same object, it will move faster . . .

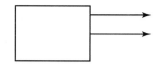

. . . or slower. . . .

Newton's (bizarre?) third law

When body #1 exerts a force on body #2,
body #2 exerts the SAME force on body #1, only opposite in direction

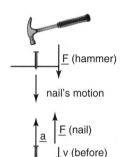

A hammer hits a nail.
The hammer exerts a force on the nail,
which penetrates inside the wall. (duh!)

But . . . The hammer stops . . .
That means there is an acceleration . . .
Which means there is an upwards force . . . .
This force comes from the nail!

We find: F(hammer) = F(nail) (magnitudes!)

We find examples everywhere:

- Person walking
- Snake moving
- Rocket propulsion
- Pushing a cart
- . . .

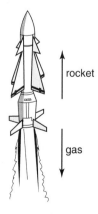

CHAPTER 5

Work and Energy

Work

Let's go back to ancient times.

Work, then, was mostly physical. Weights were moved around.

Say, you are part of the Stonehenge building crew.

You need to move a huge stone from the quarry to the construction site, say, 3 miles away.

That's a lot of work for sure.

How can I quantify it?

After a lot of thinking, people came up with the following:

$$\mathbf{W = Fd}$$

i.e., work is the force times the displacement. Block larger? More work. Quarry farther away? Also more work.

Let's be a bit more picky...

In some cases (e.g., friction), the force may be opposing motion. In this case, $\mathbf{W = -Fd}$.

Example

Suppose that you push your stalled car a distance of 150 meters by pushing on its bumper with a force of 200 newtons.

How much work is done by the force on the bumper?

The force and the distance are in the same direction, so

$W = Fd = (200\text{N})(150\text{m}) = 30,000\text{Nm}$.

I call a Nm a Joule.

Joule is indicated by J and is the unit of work.

It is the work that a force of 1N does when it moves an object by 1 m.

More examples

1. You try to push your stalled car up a shallow incline but it is too heavy.

    As a result, it rolls 150 meters *down* the hill while you push as hard as you can in the opposite (up-hill) direction.

    If you exert a force of 300 newtons the whole time, how much work does this force do on the car?

    The force is opposite to the displacement, so the work done is

    W = (−300N)(150m) = −(300)(150)Nm = −45,000 Nm.

2. How much work is done by the force which you must exert

    on a large hamburger (weight = 1 N)

    in order to pick it up off the floor?

    Assume that you raise it a distance of one meter, and that the

    hamburger moves very slowly .

    W = Fd = (1N)(1m) = 1Nm. = 1J.

    The force does one Joule of work.

Why this moving slowly business?

It just tells us that the lifting force is barely larger than the weight.

Thus, I can assume that the lifting force is equal to the weight of the object.

If the force were much larger than the weight, the block would move fast, and my approximation would not be correct.

Now, let us make a new observation

I push a block with a constant force.

The block accelerates, and its velocity increases with time.

Say, I keep pushing for a distance d.

Is there a relationship between the work done

by the force and the final velocity?

Yes.

It turns out that:

W = ½mv²

v = final velocity,

assuming initial velocity = 0

Let's see why W = Fd =½mv² makes sense.

Example

F = 10 N acts for a distance of 1m on a 1 kg mass.

What is the work?

What is the final kinetic energy?

W = Fd = 10 × 1= 10 J

$W = KE = \frac{1}{2}mv^2$.

Since $W = KE$, $KE = W = 10$ J.

What is the final velocity?

$W = \frac{1}{2}mv^2$ therefore $v^2 = 2W/m$ and $v = (2W/m)^{\frac{1}{2}}$

So, $v = (2 \times 10/1)^{\frac{1}{2}} = (20)^{\frac{1}{2}}$

Now assume you increase W. What happens to v?

It Increases.

Increase m, what happens to v?

- **Energy** is defined as the measure of a system's ability to do work.
  - We use the symbol E to represent energy.
  - Energy has the same units as work:
    - Joule
- There are various types of energy.
  - **Kinetic energy** is the energy associated with an object's motion.
    - We use the symbol KE.
  - **Potential energy** is energy associated with the system's position or orientation.
    - We use the symbol PE.

Kinetic energy is, in a sense, intuitive. The faster my car goes, the more gasoline (energy) I need to get it to and keep at that velocity. Plus, the faster I go, the more damage I will make in case of a crash.

Let's think a bit more about potential energy ....

What is potential energy?

It is the energy stored in a system that can be used to do work

Since work is related to the final velocity of an object, one way of thinking of potential energy is how fast and dangerous an object will be once the potential energy is released.

Examples
- Gravitational potential energy. A brick on top of a skyscraper. . .
- Elastic potential energy. Bow and arrow.
- Chemical potential energy. A tank of gasoline.
- Nuclear potential energy. An atomic bomb.

The total energy E of an isolated system is constant. For a mechanical system, E= K.E. + P.E.

Let's explain this.

We have seen that work can be converted into kinetic energy. Where does the work come from? From some internal source, say, an engine. The engine burns fuel to get the work done. In the beginning, we will have a full tank of gasoline, but the car will not move. The total energy of the car remains the same. Only, in the beginning the energy is stored in the tank. In the end, the energy is kinetic. In the end, the tank will be empty, but the car will be moving.

Example

For an object subject to the gravitational field of Earth,
PE = mgh.
Assume now that a block of mass 1 kg is released from rest from a height h = 5 m
And falls on the floor (where h = 0).
What is its final velocity?

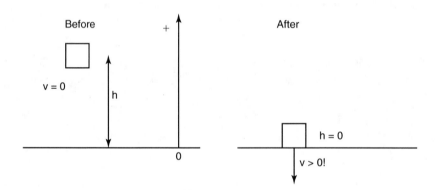

Solution

I know that PE+KE is constant.
This means that the sum of PE and KE must be the same before and after the process.
KE = ½mv², so KE(before) = 0.
KE(after) = ½mv², where v is the final velocity (which is what I need to determine).
PE = mgh, so PE(before) = mgh = 1 × 10 × 5 = 50 J.
PE(after) = 0 (h = 0!!!)
So, (PE+KE)(before) = mgh = 50 J = (PE+KE)(after) = ½mv²

$$v^2 = 50 \times \frac{2}{m} = \frac{100}{1} = 100$$
$$v = 10 \text{ m/s}$$

More important than the final velocity is to observe what happens to the energy.

Before, we have KE = 0, PE = big.

After, we have KE = big, PE = 0.

This means that we transformed potential energy into kinetic energy.

This transformation business is pretty common.

We turn

- Potential energy into electrical energy (where?)
- Atomic energy into kinetic energy
- Chemical energy into thermal energy
- Thermal energy into electrical energy
- ...

A steam engine converts *heat energy* into *mechanical energy.*

A generator converts *mechanical energy* into *electrical energy.*

An electric motor converts *electrical energy* into *mechanical energy.*

An electric stove converts *electrical energy* into *heat energy.*

A battery converts *chemical potential energy* into *electrical energy.*

Some odd things regarding kinetic energy.

$$E = \tfrac{1}{2}mv^2$$

Examples

1. If a *2000* kg car and a *10,000* kg truck are traveling at the same speed and the kinetic energy of the car is *100,000* J, what is the kinetic energy of the truck?

   First: 10,000/2000 = 5 so the truck is 5 times as massive as the car.

   Without any more calculation, the truck has 5 times the kinetic energy—

   5 times 100,000 J or 500,000 J.

This is because KE = $\tfrac{1}{2}mv^2$, so when the mass increases by 5 times, so does the kinetic energy.

2. If the kinetic energy of a car traveling *10* m/s is *100,000* J, what is the kinetic energy of the car at a speed of *30* m/s?

   Notice that *30/10 = 3* so the velocity has been multiplied by a factor of three.

   The K.E. will then be multiplied by 3^2 = 3 × 3 = 9 *times.*

   The resulting K.E. will be

   K.E. = *(9)(100,000 J)* = *900,000* J.

This is because KE = $\tfrac{1}{2}mv^2$, so when the velocity increases by three times, the kinetic energy increases by 3^2 = 9 times.

Energy is a powerful concept . . .

A roller coaster of mass m = 1 kg is at rest at point A, which has a height h = 10 m above ground. The roller coaster then starts sliding down the track, as shown in the figure. What is its final velocity when it reaches the ground at position B

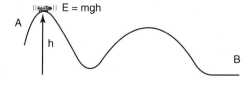

Notice that the trajectory is complicated, so cannon-ball type of approaches do not help us much here. However, we know that E = PE+KE is constant.
Initially: KE = 0, PE = mgh = 1 × 10 × 10 = 100 joule.
At point B: KE = ½mv², PE = 0
That is, I calculate the final velocity in the same way as I did when the block was falling vertically. I could not care less about the complexity of the trajectory.

Example

My car is traveling with a speed v on the highway.

I start braking the car to get to a gas station.

The brakes exert a constant force F opposite to the direction of motion of the car,

which stops at a distance d from when the braking started.

1. What is the work done by the brakes?
2. Assuming v, F and the mass of the car m to be a known value, how do I determine the stopping distance d?
3. What happens to d when I double the velocity v?

Solution: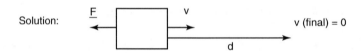

1. Work = −Fd   KE = ½ mv²
    (force, d are opposite)

2. It is: W = KE(final) − KE(initial) = 0 − ½ mv²
    So, −Fd = −½ mv², or: d = ½ mv²/F

3. If I increase the velocity by 2 times, KE goes up by 4 times.
    Thus, the work must increase by 4 times.
    Since the force is constant, the distance must increase by 4 times.

Fact

Friction opposes motion and does negative work on moving objects
—the objects lose energy. The energy lost to friction is converted into *heat energy* which is the energy associated with the disorganized motion of atoms and molecules.

Example

A 2000 kg automobile moving at 30 m/s suddenly applies its brakes and

comes to a complete stop. How much heat energy appears in the brakes?

Before stopping, the car had kinetic energy

(1/2)(2000 kg)(30 m/s)² = 900,000 J

After stopping, it still has that much energy, now converted into heat in the brakes.

CHAPTER 6

Gravitation and Planets

Let's go back to free fall . . .

Things fall with constant acceleration.

Why?

Also, we know that planets move around the Sun.

Why?

A way of understanding these things is to look at the history of

Planetary Motion

The ancient Greeks believed the sun, moon, stars and planets all revolved around the Earth.
This is called a *geocentric view* (Earthcentered) of the universe.
This view matched their observations of the sky, with the exception of the puzzling motion of the wandering planets.

Copernicus developed a model of the universe in which the planets (including Earth!) orbit the sun.
This is called a *heliocentric view* (suncentered) of the universe.
Both models made sense and had something going for them.
Careful astronomical observations were needed to determine which view of the universe was more accurate.
This task was accomplished by Tycho Brahe, who spent several years painstakingly collecting data on the precise positions of the planets.
This was before the invention of the telescope!

This was understood over a long period of time. Key were highly precise observations by Tyco Brahe. His student Johannes Kepler worked over the observations for years, and then came up with three laws. These three laws were empirical, just derived from the data. The laws were strange, and nobody really understood what they meant.

Kepler's Laws

1. Planets orbit around the Sun on elliptical orbits, with the Sun in one of the foci.

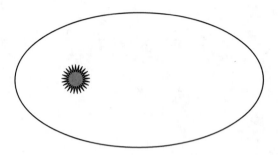

Now, this was very odd for that time, since celestial bodies were considered to be perfect. A circle was considered a perfect geometrical figure, but not an ellipse.

2. The second law was even more puzzling. Planets were found to be moving faster when closer to the Sun.

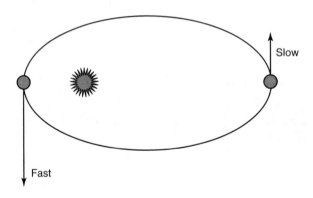

3. The third law states that the period of the orbit, T, and the radius of the orbit, r, are related by a power relation: $T^2 \propto r^3$.

Now, Kepler's laws were empirical. That is, he analyzed the data and showed that certain mathematical relationships could be derived from the data.

He did not know where these relationships were coming from, or why they did exist.

. . . Until Newton came up with his 3 laws of mechanics, plus another couple of ideas:

1. Masses attract each other.

2. The force is proportional to the product of the masses, and inversely

proportional to the square of the distance.

In formulas:

$$F = G \frac{m_1 m_2}{d^2}$$

Where G is a constant, m_1 and m_2 are the masses of the objects, and d their distance.

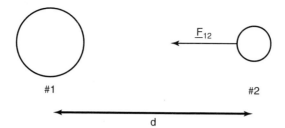

Note: force is along the line joining **the centers** of the objects. Also, d is between **the centers**, not the surfaces.

Newton proved mathematically that if

$$F = G \frac{m_1 m_2}{d^2}$$

then:

1. Only circular or elliptical orbits are possible.
2. The orbital velocity increases when planets are close to the Sun.
3. Planets that are farther away from the Sun rotate more slowly, and the ratios between period and radius are exactly what found by Kepler.

Issues with Newton's gravitation

1. Where does the force come from? Why $1/d^2$?

If the dependence were $1/d$, the Universe would be different!

Answer: We do not really know.

2. Why do not we see masses attracting each other, say, like magnets?

Answer: because G is very small.

Example

I have two masses on my desk. One mass has $m_1 = 1$kg.
The second mass has $m_2 = 1$ g. The distance between the centers is
$d = 1$ m. What is the force that the smaller mass feels?

The force is: $F = G \, m_1 m_2/d^2$,
and $G = 6.67 \times 10^{-11}$ Nm²/kg² (bizarre units by all means . . .)

So:

$F = 6.67 \times 10^{-11} \times 1 \times 10^{-3}/1^2 = 6.67 \times 10^{-14}$ N,
which is an incredibly small force.

Example

Same situation as before, this time with

$m_1 = 5.98 \times 10^{24}$ kg and R = 6370 km (aka, 6.3×10^6 m).

$F = 6.67 \times 10^{-11} \times 5.98 \times 10^{24} \times 10^{-3} / (6.3 \times 10^6)^2$

~0.01 N (or 10^{-2} N, same thing)

Still a small force, but now let us look at the acceleration of the small mass.

$a = F/m = 10^{-2}/10^{-3} = 10$ m/s²

The number should sound familiar. . . .

Why? Because we have used the mass and radius of Earth.

Conclusion

Gravitational forces become apparent when big masses are involved.

For small masses, gravitation is negligible.

Example

revisited.

We showed that Earth attracts a mass of 1 g with a force of 0.01 N.

By Newton's third law, the mass should also attract Earth.

That is, when I drop an object on the floor, the object should move towards the floor, but also the floor should raise to meet the object . . .

We do not observe this because

$a(\text{Earth}) = F/m(\text{Earth}) = 0.01/5.98 \times 10^{24} = 1.6 \times 10^{-27}$ m/s²

Which is a negligible acceleration . . .

a   F (Earth on object)

a (Earth)??   F (object on Earth)

Example

Let's calculate the acceleration due to gravity at the Moon's surface:

$$M_{\text{Moon}} = 7.22 \times 10^{22} \text{ kg}$$

$$R_{\text{Moon}} = 1.75 \times 10^6 \text{ m}$$

I know that $F = \dfrac{Gm_1m_2}{d^2} = m_2g$, the last term being the weight of my object. So:

$$g = \frac{GM}{R^2}$$

$$= \frac{(6.67 \times 10^{-11})(7.22 \times 10^{22})}{(1.75 \times 10^6)^2}$$

$$= 1.57 \text{ m/s}^2$$

$$\approx \frac{1}{6} \times g_{\text{Earth}}$$

- You would weigh one-sixth your usual weight on the Moon.
- Things fall at one-sixth the rate.
- You could jump about six times higher.

CHAPTER 7

Electricity and Magnetism

Electric charge is an intrinsic property of objects, pretty much like mass. Different from mass, however, charge comes in two flavors. Positive and negative. Elementary particles such as protons happen to have a positive charge. Other particles, such as electrons, have a negative charge. Electricity (for a change) has been known since the dawn of time. However (for a change) for a long time nobody understood it, nor thought of using electricity to do useful things.

We got to the idea of electric charge by making observations. . . .

- Hair seems to have a mind of its own when combed on a dry winter day.
- What causes the hairs to repel one another?
- Why does a piece of plastic refuse to leave your hand after you peeled it off a package?
- Why do you get a slight shock after walking across carpet and touching a light switch?

- Benjamin Franklin introduced the names **positive** and **negative** for the two types of charge.
- He also proposed that a single fluid was being transferred from one object to another during charging.

  — A positive charge resulted from a surplus of the fluid, and a negative charge resulted from a shortage of the fluid.
  — Franklin arbitrarily proposed that the charge on a glass rod when rubbed with silk be called positive.

He also showed that LIKE charges REPEL each other, UNLIKE charges attract each other
Coulomb showed that the force between charged objects is proportional to the charge, and inversely proportional to the square of the distance between the charges:

$$F = k q_1 q_2 / d^2$$

For charge, we typically use the symbol q and we measure it in Coulombs.
Where k = Coulomb's constant = 9×10^9 N m^2/C^2
d = distance, q_1, q_2 = charges on object.

Example

Two positive charges, one 2 μC and the

other 7 μC, are separated by a distance

of 20 cm. What is the magnitude of the

electrostatic force that each charge

exerts upon the other?

$$q_1 = 2\,\mu C \quad q_2 = 7\,\mu C \quad r = 20\ cm = 0.2\ m$$

$$F = \frac{kq_1q_2}{r^2}$$

$$= \frac{(9 \times 10^9\ N \cdot m^2/C^2)(2 \times 10^{-6}\ C)(7 \times 10^{-6}\ C)}{(0.2\ m)^2}$$

$$= \frac{0.126\ N \cdot m^2}{0.04\ m^2} = 3.15\ N$$

Thus, Coulomb's law is similar to Newton's law of gravitation. Because of the existence of two different types of charge (but also for other reasons), in electricity it is customary to use fields.

$$\text{electric field strength} = \frac{\text{force on a charged object}}{\text{charge on the object}}$$

- The electric field lines indicate the direction that a positive charge would move.

Below we find illustrations of electric field lines surrounding a negative and a positive charge.

- The field points towards the negative charge.
- The field points away from the positive charge.

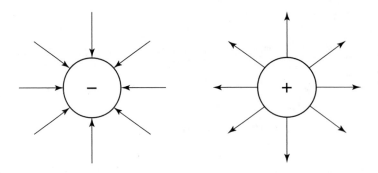

Now, electric fields (as well as point charges) exert forces on charges, which then move. When charges move, we have an electric current.

Electric currents

— The amount of charge that flows by per second is the current:

$$\text{current} = \frac{\text{charge}}{\text{time}} \longrightarrow I = \frac{q}{t}$$

— The SI unit is the *ampere* (A or amp):
- 1 amp is 1 coulomb per second.

FACT: Motion of charges does not happen without (some) fight. There are forces that oppose charge motion, and they are similar (sort of) to friction.

Resistance

- **Resistance** is a measure of the opposite to current flow.
  — Resistance is represented by R.
  — The SI units of resistance is the *ohm* (Ω).
  — A *conductor* is any substance that readily allows charge to flow through it.
  — An *insulator* is any substance through which charge does not readily flow.
  — A *semiconductor* are substances that fall between the two extremes.
- Resistance of a wire depends on:
  — *Composition*. The particular substance from which the object is made.
  — *Length*. The longer the wire, the higher the resistance.
  — *Diameter*. The thinner the wire, the higher the resistance.
  — *Temperature*. The higher the temperature, the higher the resistance.
- Resistance is similar to friction.
  — Resistance inhibits the flow of electric charge.
  — Electrons typically produce the current in metals.
  — The electrons collide with the atoms of the metal.
  — This slows them down.
  — They also lose some energy to the atoms.

FACT: A metal gets hotter when charge flows through it.

Superconductivity is a phenomenon in which the resistance of a conductor goes strictly to zero. Superconductivity is of great interest for applications. If superconducting materials were available, one could, for example, transport electric power over long distances without any losses. Alas, superconductivity requires very low temperatures, and therefore superconductors are limited to niche applications.

Voltage is the work that a charged particle can do divided by the size of the charge.
It is the energy per unit charge given to charged particles by a power supply.

$$V = \text{work}/q$$

The SI unit is the *volt* (V).
One volt equals one joule per coulomb.

Ohm's law specifies that the current in a conductor is equal to the voltage applied to it divided by the resistance:

$$I = \frac{V}{R} \quad \text{or} \quad V = IR$$

V is the voltage through the conductor,
I is the current passing through the conductor, and
R is the conductor's resistance.

How do I understand Ohm's law?

Assume you want to slide a series of blocks on a rough surface.
Since the surface is rough, you have to push the blocks, otherwise they will stop.
That is, you have to do some work on each block.

The more blocks you want to get through, the more work you have to do.

The number of blocks is "hidden" in the intensity I.

The work is "hidden" in the voltage V.

The equivalent of friction is the resistance R.

So, when I increases, V must also increase.

By the same token, when R increases, so must V.

Example

A light bulb used in a 3-volt flashlight has a resistance equal to 6 ohms. What is the current in the bulb when it is switched on?

The problem gives us:
$$V = 3 \text{ V}$$
$$R = 6 \ \Omega$$

The current through the bulb is

$$I = \frac{V}{R} = \frac{3 \text{ V}}{6 \ \Omega} = 0.5 \text{ A}.$$

Power and energy

Digression: power = energy delivered per unit time.

1 Watt = unit of power = 1 joule per second.

The **power output** of a circuit is the rate at which energy is delivered to the circuit.

Example

We just computed the current that flows in a flashlight bulb. What is the power output of the batteries?

The problem gives us:
$$V = 3 \text{ V}$$
$$R = 6 \ \Omega$$
$$I = 0.5 \text{ A}$$

So the power consumed is

$$P = IV = (0.5 \text{ A}) \, (3 \text{ V})$$
$$= 1.5 \text{ W}$$

This means the batteries supply 1.5 joules of energy each second.

Example

An electric hair dryer is rated at 1,875 watts when operating on 120 volts. What is the current flowing through it?

The problem gives us:
$$V = 120 \text{ V}$$
$$P = 1,875 \text{ W}$$

The power is given by

$$P = IV.$$

$$I = \frac{P}{V} = \frac{1{,}875 \text{ W}}{120 \text{ V}} = 15.6 \text{ A}.$$

Power and your electricity bill

- Consumption is measured in kWh. What is it?
- We can write the energy in terms of the power:

$$P = \frac{E}{t} \Rightarrow E = Pt$$

So if we know the power a circuit uses and how long the circuit operates, we can determine the energy used by the circuit.

This unit, rather mysterious, is what appears on utility bills. What is it? It is a way of expressing energy.

So: 1 kWh = 1kW × 1h = 1,000 W × 3600 s = 3,600,000 Joules.

So, utility companies charge us for the energy that we use (as one would expect), the only counterintuitive issue being the unit that they use.

Example

If the hair dryer discussed earlier is used for 3 minutes, how much energy does it use?

The problem gives us:
$$t = 180 \text{ s}$$
$$P = 1{,}875 \text{ W}$$

The energy is given in terms of power as:

$$E = Pt = (1{,}875 \text{ W})\,(180 \text{ s})$$
$$= 337{,}500 \text{ J}.$$

Magnetism

Magnets have always two poles, which we call North and South (it's just a convention). As in the case of electric charges, like poles repel each other, unlike poles attract each other.

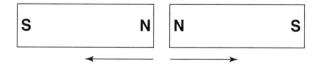

- Materials that maintain a magnetic field induced in them are called *permanent magnets*.
- As with gravitation and electrostatics, it is useful to employ the concept of a magnetic field.
- A *magnetic field* is produced by a magnet and acts as the agent of the magnetic force.
  — *Outside the magnet, the field points from north pole to south pole.*

Now, it would be nice if we could express the magnetic force as:

$$F = \text{const.} \times \text{mag}_1 \times \text{mag}_2/d^2$$

Alas, we cannot. Why? Because magnetic monopoles (the "mag"s in the equation) do not seem to exist.

Therefore, magnetism is not so simple. . . .

FACT: Electricity and magnetism are closely related.

Observation 1: Moving charges create magnetic fields. For example, a charged particle moving on a straight line will create a magnetic field in the direction perpendicular to that of motion.

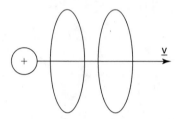

Since the field created by a single moving charge is typically very small, coils are used to create sizeable magnetic fields:

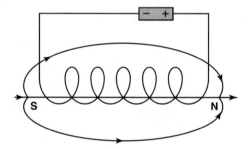

Observation 2: A magnetic field exerts a force on a moving electric charge.

- A magnetic field exerts a force on a current-carrying wire.

- However, the force experienced by a moving electric charge in a magnetic field is a bit weird. . .

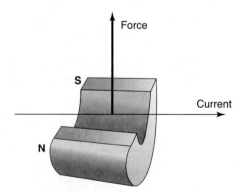

That is, the force is perpendicular to both magnetic field and current. This property of electromagnetism is exploited to make things move. In an electric motor, we pass a current through a coil. The force exerted on the coil by the magnet makes it rotate.

Observation 3: A moving magnet produces an electric field in the space around it. Equivalently, a coil of wire in motion relative to a magnet has a current induced in it.

- This process is known as **electromagnetic induction.**
- These observations allow us to introduce the **Fundamental Principle of Electromagnetism:**
  — An electric current of a changing electric field induces an electric field.
  — A changing magnetic field induces an electric field.

Now, it is possible to generate current and voltages that change with time.
Such generators are called alternated current (AC) generators, and they are pretty common.

Example: power from a wall outlet . . . is AC . . .
Power from a battery is, on the other hand, constant.
This is called direct current (DC)

More importantly, we can produce electromagnetic waves. To do that, we produce an electric field that varies with time. The electric field induces a magnetic field, which also varies with time, and induces a varying electric field, and so on. . . . The perturbation created in this way is called an electromagnetic wave and it propagates though space at a fantastic velocity. This velocity is called the speed of light, and it is indicated by the letter c. $c = 299,792,468$ m/s, or to make it simple, $c = 300,000$ km/s.

In 1888, Heinrich Hertz demonstrated the production
and detection of electromagnetic waves using spark gaps—
the first radio.

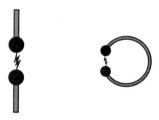

Now, Radio signals were a good thing, but generators and detectors very primitive.
The only thing that could be broadcast were sparks, long or short. Morse code was
invented out of necessity.
SOS = . . . - - - . . .

This, until the american/austrian/serbian Nikola Tesla improved the system,
and carried out the first radio transmission in St. Louis in 1893.

The idea was (copied?) expanded by Marconi,
who realized the first intercontinental transmission in 1903 between
the UK and the US.

CHAPTER 8

Light as a Troublemaker and Einstein's Special Relativity

Maxwell's theory predicts that light propagates with a speed close to 300,000 km/s.

First of, is this speed for real?

The speed which Maxwell predicted was already familiar to scientists.

In 1676, Olaf Roemer had noticed peculiar delays in the apparent orbits of Jupiter's Moons and calculated the speed of light to be close to c m/s.

So, yes, c was a number that made sense.

The problem was that light travels between Jupiter and Earth.

In between there is pretty much nothing.

Same for light reflected by the moon. What does it travel into?

This issue with propagation into vacuum was quite delicate at that time. . . .

Why?

Until and after Maxwell, people thought that waves were carried by something.
Waves can be created, and are carried by, water. No water, no waves.
Sound waves are carried by air. No air, no sound.
Some musical instruments (e.g., a guitar) work by creating waves in strings.
But e.m. waves did propagate in empty space.This was an issue. . . .

It was therefore assumed that there was a "substance" that pervaded everything, called aether.

Just as soundwaves propagate by compressing the air, light waves were supposed to propagate by distorting the aether.

The aether needed to have:

- fantastic stiffness to get the speed of light so high.
- no interaction with material things that move through it.
- no viscosity at all.

Those are pretty bold assumptions, but scientists thought it was much better than the creepy idea of waves that propagate in nothing!

To prove the existence of the aether, people thought they could measure the velocity of motion with respect to the aether. To clarify this point, we must introduce Galileo's relativity. The concept is quite simple. Consider an object moving on top of another, which is also moving in the same direction, as shown below.

An observer sitting on block #1 will say that he is moving with velocity V_1. However, for an observer at rest (#3 in the figure), the block will be moving with velocity $V_1 + V_2$.

Mathematically, we express this by saying that

V = v(object) +/− v(reference frame)

With light, it should be the same thing.

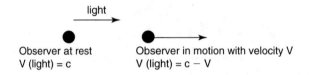

Now, people thought that the aether was at rest, and that light was moving inside the aether. Earth moves with respect to the aether, therefore one should measure a change in the velocity of light due to Earth's motion. Michelson and Morley did the experiments, and they found nothing.

Einstein's postulates of special relativity

Einstein's solution to the dilemma of the aether and the speed of light was simple and radical.

- Postulate 1: *The laws of physics are the same in any inertial frame of reference.*

- Postulate 2: *The speed of light in a vacuum is the same in any inertial frame of reference, regardless of the relative motion of the source and observer.*

The first is just a reaffirmation of the principle of relativity stated earlier.
The second is much more radical: ***light does not behave like most waves or moving objects.***

- These two postulates constitute Einstein's **special theory of relativity.**

— It is "special" because it is restricted to uniform motion.
These two postulates have some unusual implications.

1. *Events may or may not be synchronous*—Events that happen at the same time for you may not happen at the same time for someone else.
2. *Time dilation*—you observe the clock of someone moving, relative to you, to run slower than your clock.
3. *Length contraction*—you observe the length of an object moving, relative to you, to be less than an identical object at rest, relative to you AND in the direction of the object's motion.

Let's now try to understand these things.

1. **Synchronous events.** To understand this, we must first realize a thing or two about light.

Light propagates at a speed of about 3×10^8 m/s.

Remember that speed = c = space/time = s/t.

Which distance will light cover in one second?

$$c = s/t, \text{ therefore } s = ct.$$
In one second, $s = 3 \times 10^8$ m/s $\times 1$ s $= 3 \times 10^8$ m.

That is, in one second light covers about 300,000 km

In one year, light will travel a distance
$s = 9.461 \times 10^{12}$ km.
This is a HUGE distance.
Is this number of any use?

Actually, yes. Astronomers deal with large distances. They measure them in light years. When we say that an object is one light year away from us, this means that it is 9.461×10^{12} km away from us.

Fact

Proxima Centauri is the closest star to the Sun.
It is about 4 light years away from us.
This means that light emitted from the star will take 4 years to get to Earth.
The interesting part is that when we look at Proxima Centauri, we see the star as it was 4 years ago.

That is, we are getting a sort of postcard from the past.

Astronomers like distant objects because they give a snapshot on the early stages of the Universe.

Now, let us consider two lightnings that strike exactly at the same time on Earth, only, in two different places, separated by a distance D. For an observer midway between the lightnings, the two events will, indeed, happen at the same time. For an observer that is not in the center, the events will not happen at the same time, since light, say, from the lightning on the left will take a bit longer to get to the observer.

Now, let us assume that two observers are midway between the two events. One observer is at rest, the other one moves, say, to the right with a velocity v. For the person at rest, the events are synchronous. For the person in motion, events will not be synchronous. Why? It will take a time t for the light to get to this observer. In this time t, the person in motion will have traveled a distance vt to the right. Light coming from the event to the left will take a bit longer than light coming from the right. Therefore, the two events are NOT synchronous for this observer.

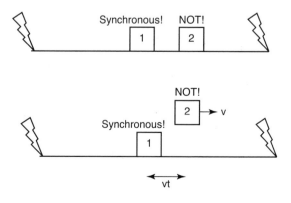

Thus, we conclude that the motion of the observer may affect the way that events are detected.

A second consequence of Einstein's postulates is that time depends on the motion of the observer. This **Time dilation** phenomenon can be demonstrated using a clock that work with light pulses:

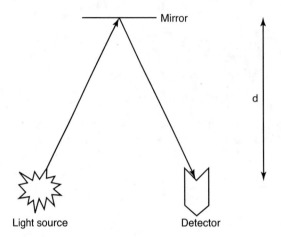

In this clock, a flash bulb emits light which bounces off a mirror and is measured by a detector.

Imagine to build two of these devices. Mount one on a spaceship moving with a speed v, and keep the other at rest.

For the clock at rest, the time between two clicks is:

$$\Delta t = \frac{2d}{c}$$

For the clock in motion, the path that the light has to cover is not 2d, but the hypothenuse of a triangle:

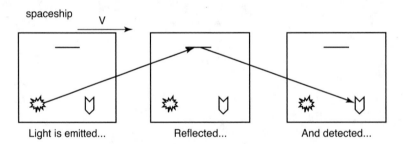

How do we calculate this new distance? We know that the beam is emitted at a time t' = 0 (t' indicates the time for the observer in motion). It will reach the detector after a time $\Delta t'$ and be reflected at a time $\Delta t'/2$.

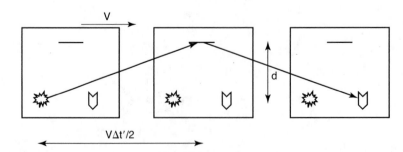

The total distance traveled by the light pulse for the observer in motion is

$$L = 2\sqrt{d^2 + (V\Delta t'/2)^2} = c\Delta t'$$

After some algebra, and remembering that $\Delta t = 2d/c$, we obtain:

$$\Delta t' = \frac{\Delta t}{\sqrt{1 - V^2/c^2}}$$

We notice that $\Delta t' > \Delta t$, that is, the clock of the observer in motion runs more slowly than the clock of the observer at rest. If you have trouble visualizing this, think this way. Say, the clock of the observer at rest has just completed one cycle (i.e., light has been emitted and detected). That will be, say, 1 second for the person at rest. Since the light of the clock in motion has to cover a larger distance, the person in motion has not detected the pulse yet (the light is still somewhere between the flash bulb and the detector). When the light is detected by the observer in motion, the clock of the observer at rest has started a new cycle already, that is, the clock in motion runs more slowly.

How significant is this effect? It is not significant for common objects and velocities (otherwise, it would have been noticed way before Einstein's time). It becomes relevant only for objects moving with $V > 0.5\ c$, which are not very common. . . .

Is this real? Yes. Detection of muons at sea level is a proof.
— Muons are subatomic particles created high in the atmosphere through collision with atmospheric molecules and energetic particles from space.
— They are created about 10 km or more above the Earth's surface.
— They have an average lifetime of 0.000 002 s.
— Since they die so quickly, they should not reach the surface in any great quantity.
— So, why do we detect such a large number at the Earth's surface?

Example

What is the mean lifetime of a muon as measured in the laboratory if it is traveling at $0.90c$ with respect to the laboratory? The mean lifetime of a muon at rest is 2.2×10^{-6} seconds.

The problem gives us:

$$v = 0.90c$$

$$\Delta t = 2.2 \times 10^{-6}\text{s}$$

So the lifetime as seen from the lab is

$$\Delta t' = \frac{\Delta t}{\sqrt{1 - v^2/c^2}} = \frac{2.2 \times 10^{-6}\text{s}}{\sqrt{1 - (0.90c)^2/c^2}}$$

$$= \frac{2.2 \times 10^{-6}\text{s}}{\sqrt{0.19}} = 5.0 \times 10^{-6}\text{s}$$

The muon "thinks" it only exists for 2.2 μs (as measured by its moving clock).

But we on the ground think the muon exists for twice as long (as measured by our stationary clocks): 5.0 μs. In 5.0 μs the muons will be able to reach Earth's surface.

Length contraction

Another phenomenon predicted by special relativity is length contraction.
How does this happen?

One method to measure the length of an object by timing how long it takes light to traverse the object.

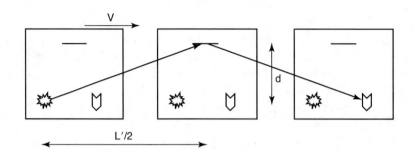

To see how this works, let us consider what the observers see after a time of one second in the frame of reference of the observer in motion. For the observer in motion the spaceship will cover a distance $L' = V \Delta t'$, and since $\Delta t' = 1$ second, $L' = V$. Since the clock of the observer in motion goes slow, for the observer at rest the time interval is longer than 1 second, so that the distance L measured by the person at rest is somewhat longer than L'.

Some math shows that

$$L' = L\sqrt{1 - V^2/c^2}$$

As for time dilation, length contraction is not observed at "normal" speeds.

Last but not least: Rest energy

Our definitions of kinetic energy and momentum no longer work in the realm of special relativity. In order for conservation of kinetic energy and momentum to hold, we must include a **rest energy, E_0,** in our calculations.
The rest energy is

$$E = mc^2$$

$E = mc^2$ means that anything that has a mass has an energy. Since c is a pretty big number, this means that macroscopic bodies contain a HUGE amount of energy.

For example, a mass of 1 kg contains

$E = 1 \times (3 \times 10^8)^2 = 9 \times 10^{16}$ Joules of energy.

It would be nice to liberate this energy. However, this is usually very difficult. Reason being that the energy is stored mainly in the nuclei. Of the atoms making up an object. And nuclei are quite stable objects. The analogy is with a bridge. Yes, should the bridge fall, a lot of energy would be liberated. But a bridge is usually a stable thing. The energy remains "frozen" as potential energy. To liberate nuclear energy, sophisticated devices such as nuclear power plants have to be built, in which not-so-stable nuclei such as Uranium's are broken up.

CHAPTER 9

Waves

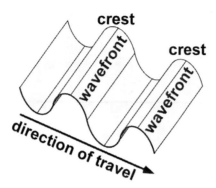

A wave is a *pattern* in the value of some quantity which is changing at every point of space.

Waves are characterized by several parameters:
amplitude, wavelength, period/frequency, and velocity

The wavelength of a wave is the distance between successive "like" points (such as crests) in a wave. Wavelengths are denoted by the greek letter λ. The amplitude of a wave is the half the distance between a trough and a crest. (Half!)

$$f = frequency$$

is the number of waves which pass a given point in one second.
The unit of frequency is inverse seconds, s^{-1}.
$1 \ s^{-1}$ is called a Hertz (Hz)

Example

Suppose that six waves pass a point in one second.

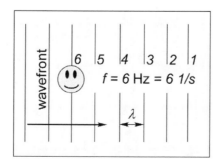

Here the waves move from left to right and the sixth wave is just passing. The frequency is
6 cycles per second or *6* Hz.

Also, waves propagate with a characteristic velocity. This velocity is given by

$$v = f\lambda$$

v = wave velocity = distance/time
Distance a wavefront travels in one second = number of waves per second times distance between wavefronts.

Examples of wave speeds
Light: 300,000 km/s
Sound: 340 m/s
Earthquakes: 2–4 km/s
Tsunamis: 600–800 km/h

FACT:

Waves do not carry matter. Example: a wave in a whip. Does the whip move? No. Do you want to be at the receiving end of the whip? No. So what does a wave transport?

A wave transports

Energy

Better definition of a wave:

- A **wave** is a traveling disturbance that transmits energy with not net movement of matter.

    The disturbance is frequently called an *oscillation* or *vibration*. The substance through which the wave travels is called the *medium*.

Let's calculate the wavelengths of some sound waves.

V(sound) ~ 340 m/s

V = fλ

So, λ = V/f

f = 20 Hz   λ = 340/20 = 17 m ~ 50 ft.

f = 1000 Hz   λ = 340/1000 = 0.34 m ~ 1 ft.

f = 20,000 Hz λ = 340/20,000 ~ 0.02 m ~ 1 in.

For electromagnetic waves, we usually talk about wavelengths, not frequencies.

The electromagnetic spectrum is here

long wavelengths   short wavelengths

Let's start with the visible.

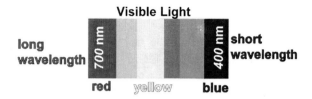

Wavelengths in this range are seen as colors. Notice that the longest visible wavelength is only about twice the shortest—a very narrow range. The visible spectrum should not be confused with the eye sensitivity to visible light. Our eye is not equally sensitive to all colors. That is, two lights bulbs, of the same power, one blue and one yellow, will be perceived by the eye as having different powers.

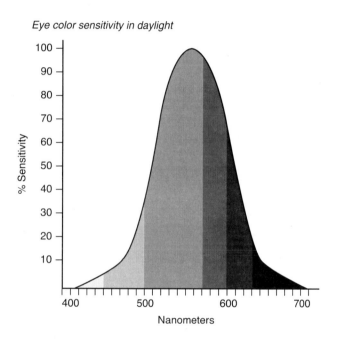

Short wavelengths: UV, X-rays and gamma rays.

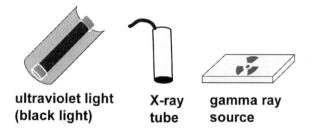

λ less than 400 nm.

That is, 400 nm down to, say, 0.01 nm and less

At these wavelengths, electromagnetic waves can knock electrons out of atoms and are referred to as *ionizing radiation*.

All these wavelengths are to be considered hazardous.

Long wavelengths: infrared and radio waves

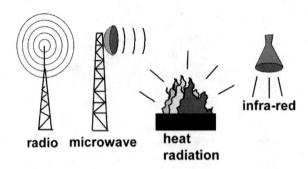

radio   microwave   heat radiation   infra-red

Note: Heat radiation is not always distinguished from the far infrared

FACTS:

- White light is a mixture of all the visible wavelengths.
- A rainbow is sunlight taken apart into its component colors.
- Most colors are actually mixtures of several different wavelengths of light.

CHAPTER 10

Optics

Fact: At the boundary between different materials, electromagnetic waves cause fluctuating electric fields which generate new waves. This fact can be exploited to generate strange effects, such as reflection, refraction, and diffusion of light.

Reflection

From a smooth surface

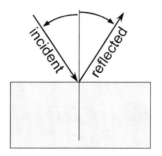

Angle of reflection equals angle of incidence.
Incident ray, reflected ray, perpendicular are all in the same plane.

Explaining mirror images:

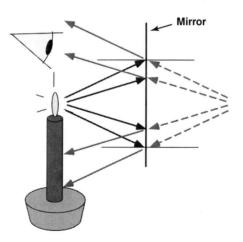

Your eye assumes that each reflected ray has traveled in a straight line.

Diffuse Reflection

Light hitting a rough surface is reflected in all directions. Such a surface can be thought of as consisting of many small-scale flat facets.

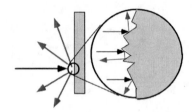

Refraction

While light in vacuum travels at velocity c, light in a medium travels at a reduced velocity, c(medium) = c/n,

Where n = index of refraction

Some numbers:
n(air) ~ 1
n(glass) ~ 1.6
n(water) ~ 1.3

The change in velocity has quite some consequences, one being that the direction of a beam changes at the boundary between two mediums.

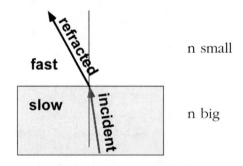

We call this process refraction.

Light in the slower region propagates at a smaller angle to the perpendicular.
(Rays and perpendicular all in the same plane.)

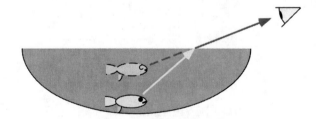

Because of refraction, the fish is not where it seems to be.
Light leaving the water is bent *away* from the perpendicular to the surface.

Total Internal Reflection

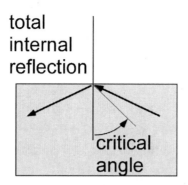

At a boundary where the speed of light changes, light coming from the "slow" side is *totally reflected* whenever its angle of incidence is greater than a *critical angle*.

Fiber optics and light pipes use total internal reflection to transport light to inaccessible places. Optical fibres are very thin light pipes which use light to transmit information over long distances.

Dispersion

The index of refraction depends on the wavelength.

$$n = n(\lambda)$$

This means that different wavelengths will be refracted at different angles.

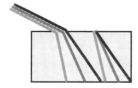

Because of dispersion, different wavelengths refract, differently. In glass, blue and violet light are slowed the most and therefore are bent more.

In a prism, the separation of different wavelengths is in the same direction at each boundary. The separations add and different wavelengths go in different directions.

Sunlight, sent through a prism splits the light into a rainbow of colors. If you look through a prism, rainbow fringes appear around light colored objects.

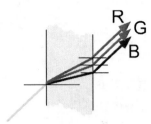

In a window pane, the separation of different wavelengths at one boundary is undone at the second boundary. Different colors are shifted sideways slightly but go in the same directions.

The slight sideways displacement of different colors is not noticeable because the directions are not changed. You do not usually see colored fringes around things when you look through a window.

Rainbow

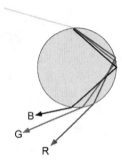

In a water droplet, light is separated by wavelength, internally reflected, and then separated still more when it comes out.

Ray approximation

Although light is really waves it is often useful to describe it in terms of *rays*—the paths of make-believe particles.

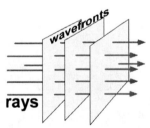

Given a family of wavefronts, draw lines perpendicular to them everywhere.
These lines are *rays*.
Light sources generally emit light in all direction.
The waves are speheres centered on the source.

The light rays are perpendicular to these expanding spherical wavefronts.

When objects are in the way of light rays, they leave shadows.

The part of the shadow where no light rays go is called the *umbra*.

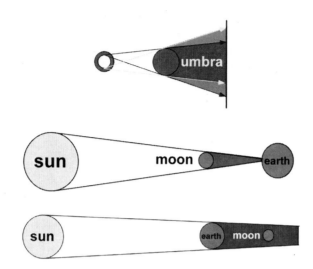

Total eclipse of the *sun:*

The umbra of the moon's shadow falls on the earth.

Total eclipse of the *moon:* The umbra of the earth's shadow falls on the moon.

Images

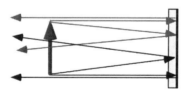

Choose some typical points on an object and follow a few rays from each object point.

The conventional object is an arrow.

Draw rays from the top and bottom of the arrow and use the laws of reflection and refraction to see where they go.

Pinhole camera

The simplest optical instrument is a pinhole camera:

A box with a hole in it and film at the back. Each ray either does or does not go through the hole.

The rays that make it through hit the film at the back of the box.

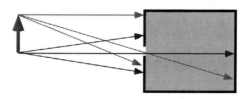

DEFINITION:

A real image is made of image points that the light rays actually go through.
If a film or screen is placed at a real image, a picture of the object will appear on it.

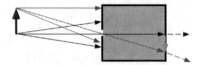

In a pinhole camera, the light from each point on the object passes through a point at the back of the camera.
A real image appears on the film at the back of the camera.

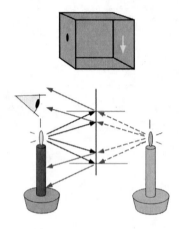

In a virtual image, the rays only appear to come from the image points but never actually reach those points.

Lenses

Lenses (and curved mirrors) bring parallel light rays to a focal point.
A converging lens is thick in the middle.
Each part of it acts like a prism, bending light toward the axis.
Rays parallel to the axis all pass through a *focal point*.

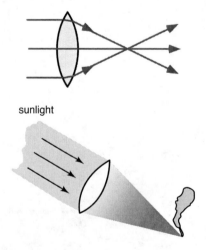

sunlight

A converging lens will concentrate sunlight and start a fire.

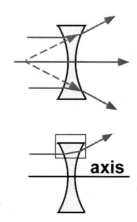

A diverging lens is thin in the middle.
Each part of it acts like a prism, bending light away from the axis.

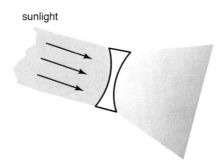

A diverging lens spreads sunlight out and cannot start a fire.

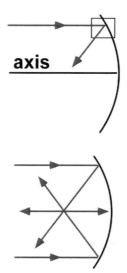

A concave mirror reflects light toward the axis.
Rays parallel to the axis are all reflected through a *focal point*.

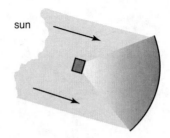

A solar furnace uses a large concave mirror to concentrate sunlight into a small chamber. The temperature can reach 6000 degrees C.

Some more Facts on Lenses

In lenses, there are two points which are a bit special. These points are called focal points.

Fact

When an object is in one of the focal points, rays exit the lens parallel to each other.

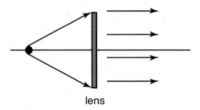

Conversely, parallel rays are focussed at the focal point.

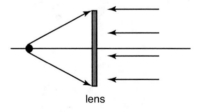

These properties of the focal points can be exploited to fabricate devices.

The Projector

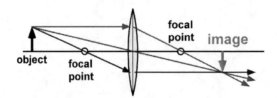

To project an image, we place the object outside the focal point.

Look at what happens when the object is moved. . . .

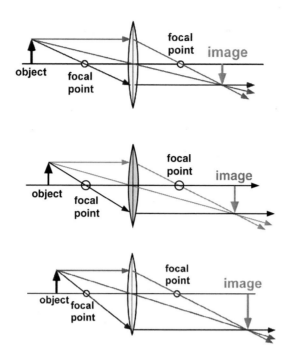

The Camera

In a camera, replacing the pinhole by a lens puts a lot more light into the image. But now the film has to be at the right distance from the lens—the camera must be focused and objects at different! distance will not all give sharp images on the film.

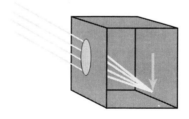

The Magnifier

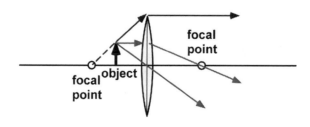

In a magnifier, the object is inside the focal point.

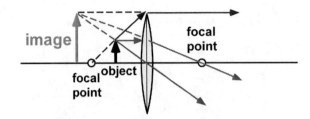

In a magnifier, the image is virtual

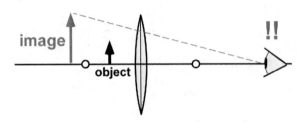

More on the magnifier

Magnifier close to the object: Little magnification.

Magnifier farther from the object: More magnification.

Magnifier still farther from the object: Still more magnification.

Magnifier at almost focal distance from the object: Distorted image due to defects in lens.

Object at or beyond focal point: No image. Colors from chromatic aberration.

The Eye

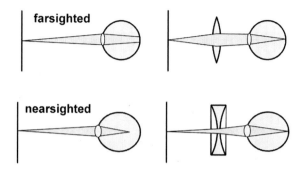

In the eye, a lens focuses images on the retina. If the lens lacks converging power or the retina is too close to the lens, then rays from nearby objects are not brought to a focus. A converging lens corrects the problem. If the lens has too much converging power or the eyeball, is too long, rays from distant objects focus before they reach the retina. A diverging lens corrects the problem.

Farsighted people wear glasses with converging lenses.
The lenses have thin edges and the wearer's eyes are magnified.
Nearsighted people wear glasses with diverging lenses.
These lenses have thick edges and make the wearer's eyes look smaller.

The eye, line any camera, consists of a lens to form an image, a diaphragm to regulate light, and a detector.

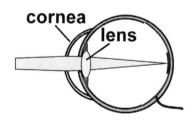

FACT: Light is focused partly by the cornea and partly by the lens.

Surgery to correct nearsightedness changes the shape of the cornea. making it flatter.
Sometimes the lens becomes cloudy—a cataract.
Surgery to correct a cataract removes the lens.
If a new lens is not implanted, then very thick converging-lens glasses are needed to bring images to a focus on the retina.

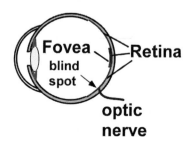

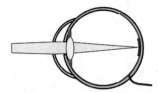

To focus the light from objects *at a distance*, muscles pull on the lens and flatten it.

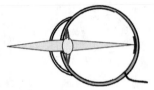

To focus the light from objects *nearby*, the muscles relax and the lens becomes thicker.

The lens grows throughout life until, at around age 45, the focusing muscles become ineffective. Bifocal glasses have segments which consist of lenses with different converging power. The lost focusing ability is then replaced by looking through one segment or another.

The **iris** helps regulating the amount of light that enters the eye.

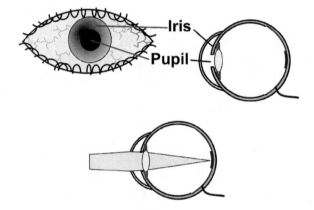

The eyedrops used during eye examinations paralyze the iris and leave the pupil wide open. Walking out of an eye exam into a sunny day is extremely painful because the eye has lost its defense against bright light.

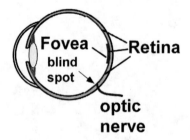

A layer of specialized nerve cells, called the *retina* detects the light. Where the cells connect to the *optic nerve,* there is a *blind spot.*

Optical illusions

The brain receives only a summary of the information that gets into the eye. What we think to see is actually a reconstruction of the brain, based on certain assumptions.

Visit the website

http://www.davidairey.com/images/design/Escher-relativity.jpg to get examples of optical illusions.

We will now briefly describe how optical illusions come about.

The problems start in the eye. The information provided by the cells in the retina. Is way larger than the transmission capacity of the optical nerve. Sophisticated compression algorithms are implemented already within. The eye to transmit the information. This has somewhat bizarre consequences:

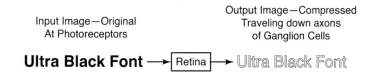

Also, some systems of nerve cells detect specific angles between edges, others detect straight lines, and so on.

Near the edges of the field of vision, only *changes* in the image are detected.

Because the brain receives a summary of local relationships, it sometimes fails to reconstruct global relationships correctly. Objects in different parts of a picture can look different even though they are identical because their relationships to nearby objects are different.

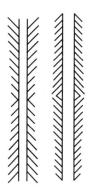

Local clues about the direction and spacing of lines mislead the eye.

Illusions reveal how the brain processes visual information.

More Websites with Optical Illusions:

http://www.michaelbach.de/ot/ang_hering/index.html
http://www.michaelbach.de/ot/ang_poggendorff/index.html
http://www.michaelbach.de/ot/ang_cafewall/index.html
http://home.comcast.net/~eschermc/Ascending_and_Descending.jpg

CHAPTER 11

Quantum Physics

The quantum hypothesis

By the end of the 1800's, physics had made significant progress. Some physicists feared "that *all* of their questions might soon be answered."

But some problems defied solution:
- blackbody radiation,
- photoelectric effect.
- atomic spectra.

Blackbody radiation

- Everything around you is constantly emitting electromagnetic (EM) radiation.
- A perfectly "black" body would:
  — absorb all light and other EM radiation incident upon it, and
  — be a perfect emitter of EM radiation.
- The EM radiation emitted by it is called *blackbody radiation* (BBR).

The radiation emitted by a black body is a mixture of several different frequencies, and it depends on the temperature of the body. For example, fresh lava emitted from a volcano is red; when the lava cools, it emits heat, i.e., infrared radiation. The Sun has a temperature of 5000 K, and it is yellow. A good representation of blackbody radiation can be found at: http:// lectureonline.cl.msu.edu/~mmp/applist/blackbody/black.htm

The principles of electromagnetism explain some of this:

A blackbody emits radiation since the atoms and molecules in the body are continually oscillating.

Recall that a vibrating electric charge emits EM radiation.

However, some things could not be answered, for example, the relation between temperature and color.

A "solution"was devised in 1900 by Planck.

He developed a mathematical equation that accurately fit the blackbody radiation curve. Planck's formula reproduced the data, but gave no physical insight. He then developed a model that would produce the desired equation.

He proposed that an oscillating atom in a blackbody can only exchange certain fixed values of energy.

It can have zero energy, or a particular energy E, or $2E$, or $3E$, ...

- This means that the energy of each atomic oscillator is **quantized**.
- The energy E is called the fundamental **quantum** of energy for the oscillator.

Quantization may sound like a big word, but we see it every day.

Buy gas from a gas station: you can buy any amount you want, e.g., 1 gal, 10 gal, 10.5 gal. . . . This is not a quantized system.

Buy milk at a supermarket. Can you buy 1.3 gal?

Milk is sold by supermarkets in a quantized way, i.e., you can only buy multiples of a certain quantity.

Planck determined that the basic quantum of energy is proportional to the oscillator's frequency:

$$E = hf \quad h = 6.63 \times 10^{-34} \, J \times s$$

The constant h is called *Planck's constant*.

The allowed energies are then

$$E = 0, \quad \text{or } E = hf, \quad \text{or } E = 2hf, \ldots$$

The Photoelectric Effect.

. . Was another puzzling thing. . . .

Scientists saw electrons being emitted by metals illuminated with a suitable wavelength.

Experiments also showed that

- For every metal, there is a threshold frequency
- Light with frequency below threshold will do nothing, no matter how long we wait.
- Light above the threshold frequency induces electron emission, no matter how weak the light source is.
- There is no time delay between the illumination and the emission of electrons.

This was quite a headache . . .

Scientists thought of the lines of Wave theory, and envisaged light as an oscillating electric field.
In this view, only the electric field strength should affect electrons.
Prediction: Sufficiently intense light should eject electrons no matter what the frequency.

This prediction was wrong, since light below a certain frequency does not generate electrons, no matter what its intensity is.

There was more in store. . . .

Wave Theory: The power in a light beam is spread out.
The area near an electron intercepts very little.
Prediction: It should take some time for an electron to absorb enough energy to be ejected.

Alas, no delay between illumination and electron emission was observed.

The key to understanding the photoelectric effect is to think of light as a bullet.

Say, we have a projectile with energy E which collides with a target. The target is glued to some support, and it takes an energy w to break the glue.

If all the incident energy is transferred to the target, it will fly off with kinetic energy KE = E $-$ w

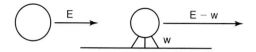

This assumption explains all the oddities of the photoelectric effect.

This means that electromagnetic waves can also behave as particles. We call these particles photons, and they have an energy E = hf. h = Planck's constant.

In the photon theory, greater light intensity simply means more photons. Each photon ejects electrons in the same way, so more intensity means more electrons.

Threshold frequency and electron K.E. are not affected by intensity.

Atomic spectra

- The third problem that classical physics could not resolve is the emission spectra of the elements.
- Consider light emitted by a tungsten filament. It is more or less white, that is, it contains all colors.
- Consider now the Na lights on a highway. They are yellow, although there is no filter to make them look yellow.
- **WHY?**

Bohr model of the atom

- Bohr constructed a model of the atom called the **Bohr model**
  - The atom is a miniature "solar system."
    - the nucleus is at the center and the electrons move about the nucleus in well-defined orbits.
  - The electron orbits are quantized.
    - the electrons can only be in certain orbits about a given atomic nucleus.
  - Electrons may "jump" from one orbit to another.

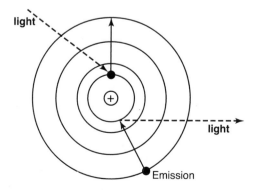

Based on the energy differences between orbits we can draw energy diagrams.

The lowest orbit of an electron is called a ground state.

Higher orbits have higher energies.

The energy differences ΔE between orbits is the energy that we need to supply to move an electron between orbits.

If we supply anything less than ΔE, nothing happens.

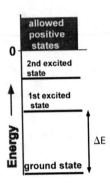

If an electron is brought to an excited state, say by a collision or by absorption of light, it will go back to the ground state and emit light of a specific wavelength. This is what happens with the yellow sodium lamps on highways. By the same token, if yellow light crosses a cloud of sodium vapors, it will be absorbed, since the electrons will go from the ground state to the excited state. This latter observation explains the presence of dark lines (Fraunhofer lines) in the Sun's spectrum, which can be found at http://www2.jpl.nasa.gov/basics/bsf6-3.php. The lines are due to a layer of rarefied gases that surround the Sun. The Sun itself emits a continuum of wavelengths, some of which are absorbed by the gases in the outer layers:

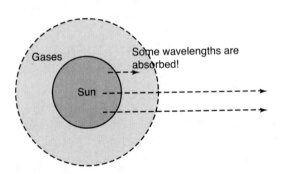

So, we have seen that

- Black Body
- Photoelectric Effect
- Absorption and Emission Spectra

Can only be explained based admitting that Atoms can exchange only discrete amount of energy. We call these energy packets quanta.

The Bohr model with its quantized orbits helps explaining these effects, but it is not the whole story.

The whole story is in the Schroedinger Equation and in the Heisenberg indetermination principle. They are as important as F = ma and E = mc², and an array of effects are derived from them.

Schroedinger's equation:

$$\frac{h^2}{2m}\left[\frac{\partial^2\psi}{\partial x^2} + \frac{\partial^2\psi}{\partial y^2} + \frac{\partial^2\psi}{\partial z^2}\right] + V\psi = ih\frac{\partial\psi}{\partial t}$$

Selected consequences of quantum mechanics:

1. We cannot localize particles with an absolute precision

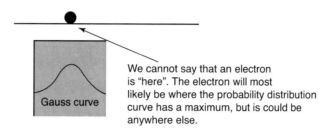

We cannot say that an electron is "here". The electron will most likely be where the probability distribution curve has a maximum, but is could be anywhere else.

2. Tunnel effect

Electrons and other small particles can tunnel through a potential energy barrier. It is even theoretically possible (but extremely unlikely) for a tennis ball to get through a wall.

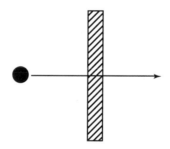